TOO MANY KANGAROO THINGS TO DO!

by Stuart J. Murphy

illustrated by Kevin O'Malley

HarperCollinsPublishers

HarperCollins®, ☰®, and MathStart™ are trademarks of HarperCollins Publishers Inc. For more information about the MathStart Series, please write to HarperCollins Children's Books, 10 East 53rd Street, New York, NY 10022

Bugs incorporated in the MathStart series design were painted by Jon Buller.

Library of Congress Cataloging-in-Publication Data
Murphy, Stuart J., date.
    Too many kangaroo things to do! / by Stuart J. Murphy ; illustrated by Kevin O'Malley.
        p.   cm. — (MathStart.)
    "Level 3."
    Summary: A surprise birthday party planned for a kangaroo by his friends provides many opportunities for the reader to add and multiply various things.
    ISBN 0-06-025883-7. — ISBN 0-06-025884-5 (lib. bdg.) — ISBN 0-06-446712-0 (pbk.)
    1. Multiplication—Juvenile literature.   2. Addition—Juvenile literature.   [1. Multiplication.
2. Addition.]   I. O'Malley, Kevin, date, ill.   II. Title.   III. Series.
QA115.M88   1996   95-20879
513.2'11—dc20   CIP
                                                                                          AC

Typography by Tom Starace
2   3   4   5   6   7   8   9   10
❖

To Nick—
*who always has
too many Nick things
to do.*
—S.J.M.

To Noah,
*Number 2 in the
pecking order but
right at the top
of my heart.*
—K.O.

"Hi, Emu! It's my birthday. Will you play with me?" asked Kangaroo.

"Sorry, Kangaroo, I have too many emu things to do."

5

6

"I have to bake one cake,
spread two colors of frosting,
decorate the cake with three flowers,
and add four big candles."

1 emu

1 x 1 cake = **1** cake

1 x 2 colors of frosting = **2** colors of frosting

1 x 3 flowers = **3** flowers

1 x 4 candles = **4** candles

That's **10** emu things!

"I guess I'll hop down to the river.

Maybe the two platypuses will play with me."

"Hey there, Platypuses! Do you want to play with me?
Today's my birthday," said Kangaroo.

"Sorry, Kangaroo, we have too many platypus things to do."

"We each have to slice one kiwi,

squeeze two oranges,

pour three cans of ginger ale,

and scoop four big scoops of sherbet."

2 platypuses

2 **x** 1 kiwi = **2** kiwis

 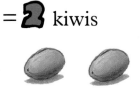

2 **x** 2 oranges = **4** oranges

2 **x** 3 cans of ginger ale = **6** cans of ginger ale

2 **x** 4 scoops of sherbet = **8** scoops of sherbet

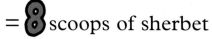 

That's **20** platypus things!

"I guess I'll hop over to the eucalyptus tree.
Maybe the three koalas will play with me."

"Hi, Koalas! Today is my birthday.
Won't you play with me?" asked Kangaroo.

"Sorry, Kangaroo, we have too many koala things to do."

"We each have to find one box,
wrap it with two sheets of wrapping paper,
tape it with three pieces of tape,
and tie four ribbons into a big bow."

3 koalas

3 x 1 box = **3** boxes

3 x 2 sheets of wrapping paper = **6** sheets of wrapping paper

3 x 3 pieces of tape = **9** pieces of tape

3 x 4 ribbons = **12** ribbons

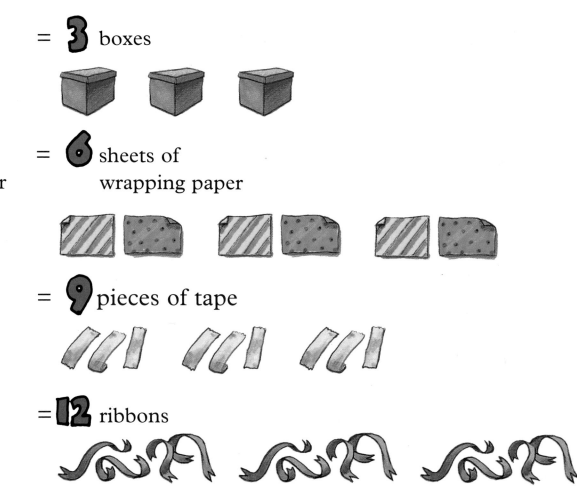

That's **30** koala things!

"I guess I'll hop up to the cave.
The four dingoes live there.
Maybe they will play with me."

"Hello, Dingoes! Would you like to play with me? Today is my birthday," said Kangaroo.

"Sorry, Kangaroo, we have too many dingo things to do."

23

"We each have to plan one game,
hang two streamers,
make three prizes,
and blow up four big balloons."

4 dingoes

4 × 1 game = **4** games

4 × 2 streamers = **8** streamers

4 × 3 prizes = **12** prizes

4 × 4 balloons = **16** balloons

That's **40** dingo things!

"All the other animals have too many things to do.

No one will play with Kangaroo."

"Hey, Kangaroo, come on back. We have a surprise for you!"

29

"HAPPY BIRTHDAY!

10 emu things + 20 platypus things + 30 koala things +

30

We really had too many *kangaroo* things to do!"

40 dingo things = 100 KANGAROO THINGS!

31

If you would like to have more fun with the math concepts presented in *Too Many Kangaroo Things to Do!*, here are a few suggestions:

• Read the story together and ask the child to describe what is going on in each picture.

• Ask questions throughout the story, such as "If one emu bakes one cake, how many cakes are baked?" or "If two platypuses squeeze two oranges each, how many oranges have been squeezed?"

• Review the math summary pages and encourage the child to talk about the number of activities each animal has to do and the total number of activities that have to be done.

• Together draw some animals—one bear, two bunnies, three cats, four dogs—and imagine they are planning a birthday party. Let the child assign each set of animals a number of tasks. "How many bunny things are there to do?" "How many dog things?"

• Gather some building blocks, dolls, or trucks. Arrange the items in even rows on the floor—for example, 5 rows of 3 dolls each, 2 rows of 4 blocks each. How many are there in all? If you have 12 toys, how many ways can you arrange them?

• Look at things in the real world—shoes, tricycle wheels, and table legs—and help the child make a list of items that come in twos, threes, and fours. Ask questions like "If we have three tables, how many table legs do we have?"

Following are some activities that will help you extend the concepts presented in *Too Many Kangaroo Things to Do!* into a child's everyday life.

Cooking: Make some cookies. Arrange the dough on each baking sheet differently—for example in rows of twos, threes, and fours. While they bake, practice multiplying. Ask: "How many cookies are on the sheet?" and "After three sheets, how many cookies will we have?"

Nature: Draw lots of insects with different numbers of legs—spiders, ants, beetles, mosquitoes, caterpillars, etc.—on cards. Count the number of legs on each insect. How many legs are there if you have two spiders, three ants, or four caterpillars?

Money: Set up a snack stand. Find a variety of treats to sell—cups of lemonade, cookies, apples, etc.—and price each kind of food differently. Use multiplication to figure out how much money each customer owes.

The following books include some of the same concepts that are presented in *Too Many Kangaroo Things to Do!*:

- ANNO'S MYSTERIOUS MULTIPLYING JAR by Masaichiro and Mitsumasa Anno
- THE 12 CIRCUS RINGS by Seymour Chwast
- EACH ORANGE HAS 8 SLICES by Paul Giganti, Jr.